Backyard Butterflies

of the
Lowcountry

by Judy Drew Fairchild

©2024
Nature Walks with Judy, LLC

Nature Walks with Judy
It's all connected!

Author's note:

I am not a scientist, and this is not a field guide. While I love field guides, I always want to know more about the organisms I am looking up. What does it eat? Who eats it? Can I create a habitat to encourage it? When I had small kids, I wanted a book full of photos I could use to learn and share what all developmental phases of an organism looked like. For the last 20 years, I've been using my camera to learn and remember the fascinating creatures around me.

Caring for our planet means rethinking our relationship with insects. Butterflies are a great place to start because they are so captivating. Think of this as an introduction; a handbook for creating and understanding your insect-friendly yard. A healthy insect population leads to other wildlife arriving: a single pair of chickadees might feed their young more than 6,000 caterpillars per clutch.

So I've put this book together to encourage you to get outside and look for these butterflies. I'm going for a beginner-friendly vibe, so I use more common names than scientific names. Obviously, I can't include everything that flies or crawls through your yard, but once you get to know a handful, it's easier to learn new ones. And if you know which native plants provide food and shelter for the butterflies, even better! You'll find lots of resources on pages 80–97.

A huge shoutout of thanks to the Master Naturalist and iNaturalist communities, who have helped me get to know (and show) the butterflies, moths, and plants found in this book. Nature creates the most amazing network of connections, and we are just beginning to comprehend how vast that connected world is.

At the end of this book (and on our website at https://naturewalkswithjudy.com) you'll find sources and links to online videos that show lots of butterflies in action, and other nature, too,

Judy Drew Fairchild is a nature educator and South Carolina Master Naturalist who lives on Dewees Island, a nature preserve near Charleston, SC, where gas-powered vehicles and broad-spectrum pesticides are not allowed, there are no lawns, and only native plants can be planted in the ground.

Table of Contents

Lepidoptera: An Introduction

Butterflies and moths are in the family Lepidoptera, which is Greek for "scaled wing." This close-up of a Palamedes Swallowtail wing shows the scales, like tiny layered hairs which give the wings their colors by bending or scattering light. Scales may also assist with giving the butterfly more lift.

Parts of a Butterfly

Like all insects, butterflies have three main body parts: a **head**, a **thorax**, and an **abdomen**. The head has compound eyes, antennae, and a tube-like tongue called a **proboscis**. The jointed legs and two pairs of wings are joined to the thorax. The front wings are called forewings, and the back wings are hind wings. The abdomen has the reproductive tract and digestive organs.

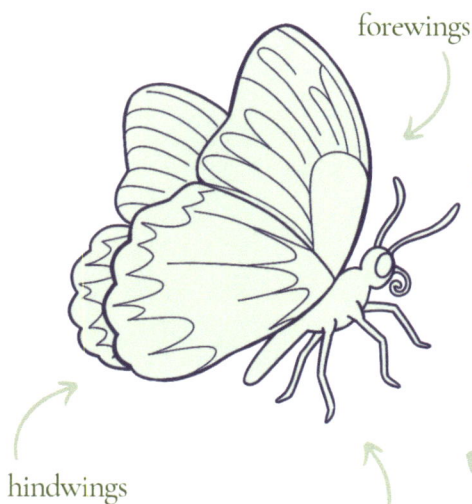

forewings

hindwings

The abdomen contains the digestive tract and reproductive organs.

The antennae can help with balance, and detect smells, wind speed and direction. They are crucial for determining direction for migration.

The thorax (the middle section) has strong muscles to move the jointed legs and both pairs of wings.

The legs have chemical receptors that help the butterfly know which plants are good for food or laying eggs.

The proboscis can work like a straw, but butterflies can also use it to send fluid out to dissolve salts and minerals in order to ingest them. It's specially modified to get nectar from plants, and can be rolled up or extended to get nectar from flowers.

Butterfly Families

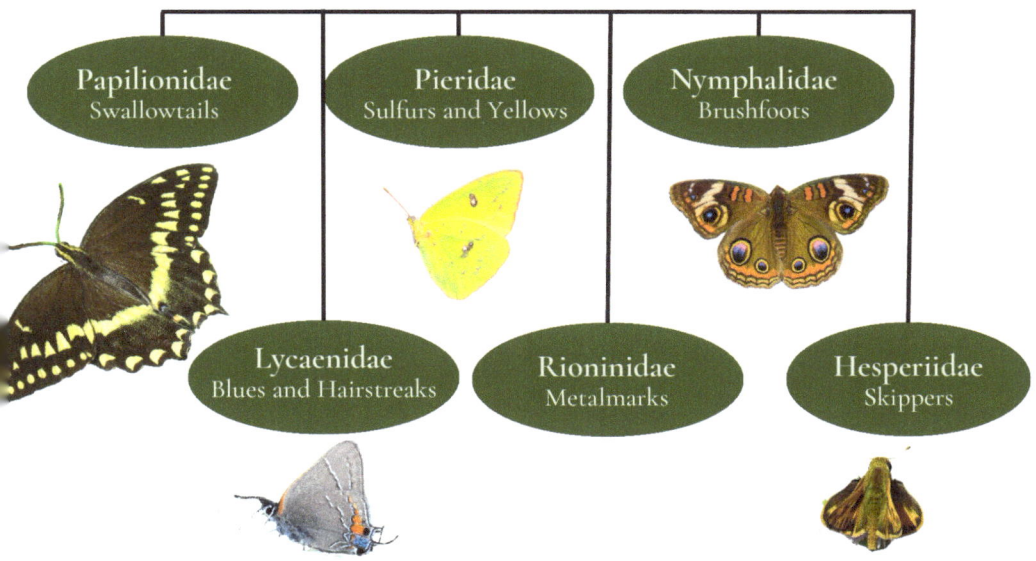

Papilionidae
Swallowtails

Pieridae
Sulfurs and Yellows

Nymphalidae
Brushfoots

Lycaenidae
Blues and Hairstreaks

Rioninidae
Metalmarks

Hesperiidae
Skippers

You don't really need to know the butterfly families, but it can some in handy to know the characteristics if you find a butterfly you can't identify. It also gives us an easy way to organize butterflies. (I don't have metalmarks in my yard, so there aren't any in this book!)

Since my goal is really to get you to connect with nature more deeply than is possible with just a field guide, I don't cover *all* the possible butterflies you'll find, but there are some suggestions for field guides at the end.

Butterfly Life Cycles and Stages

Like all insects, butterflies wear their skeletons on the outside of their bodies. Even squishy caterpillars have a sort of exoskeleton: they might have spiny spikes, or they might have knobby bumps, poisonous chemicals, etc. They all undergo some sort of metamorphosis where they change into something that seems COMPLETELY DIFFERENT!!!

Butterflies have four life stages: egg, caterpillar, pupa, and adult.

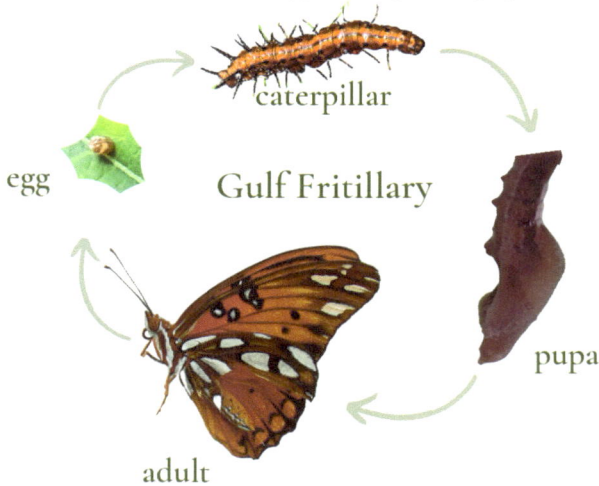

caterpillar

egg

Gulf Fritillary

pupa

adult

Adult butterflies lay eggs, which hatch into caterpillars (larvae).

Within the caterpillar stage, butterflies have several "instars," or growth phases. As the caterpillar grows, it sheds its skin, changing its appearance as it grows.

Imperial moth larva showing those strong jaws and claspers

Because the main job of a caterpillar is eating, they have strong jaws and claspers to hold on to leaves.

The larva (caterpillar) becomes a pupa (chrysalis).

The caterpillar stops eating and searches for an appropriate spot for pupation. Using silk spinnarets (like spiders do), it attaches itself to a surface.

It sheds its skin a final time, and forms a chrysalis. In this **pupal stage**, the caterpiller is known as the **pupa**.

The pupa (chrysalis) ecloses, and a butterfly emerges.

Depending on the species and the time of the year, it can take anywhere from a week to several months for the butterfly to emerge from the chrysalis.

Eventually, the chrysalis becomes almost transparent and shows the colors of the butterfly wings, and then it splits! The butterfly emerges head downward, and quickly flips right side up to allow gravity to assist in moving blood from the swollen thorax out to the wings. This procedure of emerging is called **eclosing**.

The butterfly rests until the veins in the wings fill and it is strong enough to fly.

Dangers and Defenses

Butterflies have a wide variety of tricks and adaptations to help them evade predators.

As caterpillars eat plants, they turn sunlight (photosynthesized energy) into a digestible nugget of protein and fat. This makes them a great source of nutrition for a lot of predators. Other insects like assassin bugs eat them, and birds spend a lot of time looking for tiny caterpillars or even pupae to bring back to their nests of hungry baby birds. Lizards, frogs and spiders all lie in wait to grab adult butterflies.

Adaptations:

aposematism

When an organism advertises that it is unsafe to eat, using bright colors. Monarchs caterpillars and butterflies both do this.

camouflage

Some butterflies can blend in completely with the leaves around them. Chrysalises and caterpillars blend with leaves.

concealment

Some butterflies hide in parts of the host plant leaves and flowers: even sewing themselves in with silk!

deception

Some butterflies have antenna-like tail filaments and fake eyespots on their tails so a predator will bite the tail rather than the head. Others have large eye spots that might confuse predators into overestimating their size.

mimicry

Some butterflies taste horrible to birds; others mimic the bad-tasting butterflies to protect themselves from predators. Many butterflies and caterpillars have large eye spots to confuse predators into overestimating their size.

offenses

Some caterpillars have spiny exoskeletons, some have hairy filaments that sting, some become toxic by ingesting plant toxins, some emit a stinky scent.

Lowcountry Butterflies

1. Gulf Fritillary
2. Phaon Crescent
3. Silver-spotted Skipper
4. Checkered Skipper
5. Falcate Orangetip
6. White M. Hairstreak
7. Painted Lady
8. Common Buckeye
9. Black Swallowtail
10. Variegated Fritillary
11. Eufala Skipper
12. Long-tailed Skipper
13. Gray Hairstreak
14. Viceroy
15. Pipevine Swallowtail
16. American Lady
17. White Peacock
18. Little Wood Satyr
19. Giant Swallowtail
20. Mourning Cloak
21. Palamedes Swallowtail
22. Fiery Skipper
23. Monarch
24. Eastern Tiger
    Swallowtail
25. Queen
26. Red Admiral
27. Zebra Longwing
28. Horace's Duskywing
29. Cloudless Sulfur
30. Sleepy Orange
31. Little Yellow
32. Eastern Pygmy Blue

Papilionidae: The Swallowtails

Swallowtails are some of our largest butterflies. They tend to fly gracefully, sailing on the wind currents from place to place. They rest with their wings open or closed.

All of our swallowtails have small extensions on their hindwings that look like tails. This is how they get their name.

Swallowtail caterpillars are tiny masters of deception. Most begin life looking like bird poop on leaves.

Most swallowtails lay their eggs one at a time on the leaves of the host.

As the caterpillars grow, the final instar looks like it has a face, which could mislead a predator into thinking it is a frog or young snake: perhaps a bird or lizard will move on.

If deception isn't enough to warn off predators, these caterpillars have a specialized defensive organ called an osmeterium, which is extended (everted) when the caterpillar feels threatened, It looks a little like a snake's tongue, and it can even release a foul smell!!!

Palamedes Swallowtail

Green Tree Frog

Eastern Tiger Swallowtail

- striped body
- mostly yellow
- dark morph is dark gray
- "tiger" stripes on tops and bottoms of wings
- innermost spots are blue

3.12 - 5.5"

Black Swallowtail

- spotted body
- more black than brown
- lower spots separated
- smaller (2.5-4.25")

Giant Swallowtail

- yellow body
- much larger
- upper spots fused into stripes on back
- underside mostly yellow
- largest (4-6")

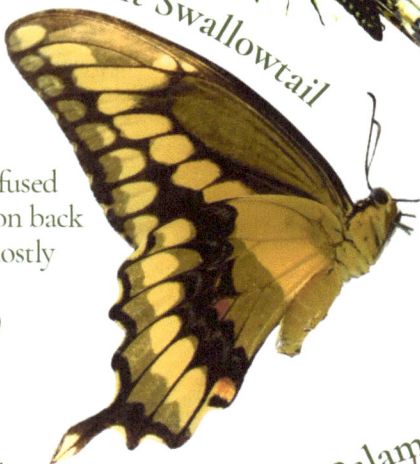

Pipevine Swallowtail

- spotted body
- purplish hindwings
- feeds on noxious pipevine
- 7 orange spots on hindwing
- 3-3.5"

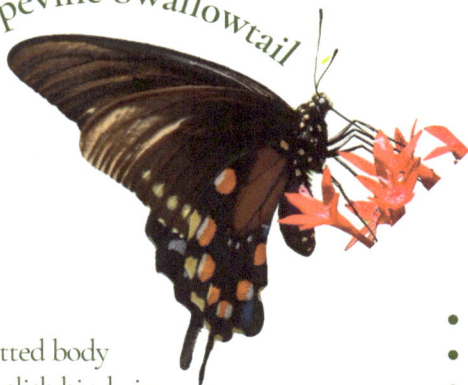

Palamedes Swallowtail

- striped body
- browner spots fused into a stripe
- yellow stripe on wing parallel to body
- innermost spots are orange
- larger than black (4-5.5")

Giant swallowtail butterflies are the largest butterflies in the Lowcountry.

Their undersides are mostly yellow, with the spots fused into stripes. The body is yellow. This butterfly seems to sail with leisure from plant to plant.

Giant Swallowtail

Their upper side is black with prominent yellow stripes that make a strong triangle across the back and extend down the wings.

Host plants include members of the citrus family, as well as toothache trees (*Zanthozylem clava-herculis*) pictured here.

Giant swallowtails will nectar on a wide variety of plants, including blue flag iris.

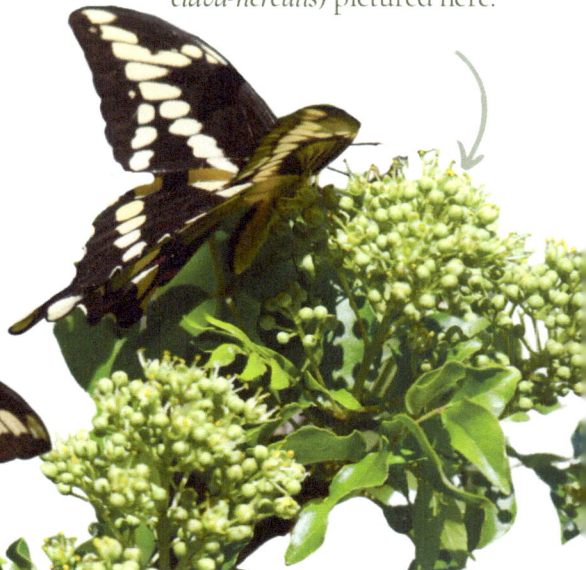

The caterpillar goes through several instars. In the final phase, it looks like it has a face, which might fool a bird into thinking it's a snake. If that doesn't work, these caterpillars can **evert** their **osmetarium** to care off predators.

When the caterpillar is ready to pupate, it forms a sort of J shape and attaches to a twig or surface with spun silk, and forms a chrysalis.

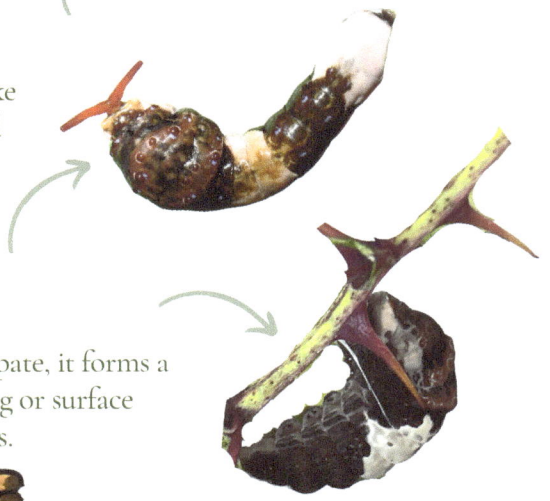

The chrysalis looks like a spur or part of the original plant.

Papilio cresphontes
4 - 6"

Palamedes Swallowtail

With striped bodies and wings of a rich chocolate brown, Palamedes swallowtails have yellow, orange, and blue spots. The spots on the top of the hindwing are fused together. On the underside, a yellow stripe on the hindwing parallels the stripes on the body.

They're larger than black swallowtails and smaller than giant swallowtails.

Palamedes swallowtails will nectar on a wide variety of plants, but prefer Red bay, Swamp bay, and even Sassafras as host plants.

To attract these butterflies, try planting Red Bay trees. (*Persea Borbonia*)

Early instars look similar to giant swallowtail caterpillars, but the final phase is green, and looks a little like a frog.

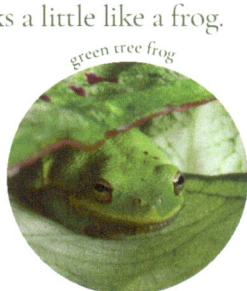

green tree frog

Palamedes caterpillar

When the caterpillar is ready to pupate, it turns yellow and stops eating, finding an appropriate perch to hang in an inverted J-shape.

The pupa, or chrysalis, might be green or brown, and has two "horns." It is attached with tiny silk anchors.

Papilio palamedes
4-5½"

Black Swallowtail

Black swallowtails are smaller than giant and Palamedes swallowtails. They are one of the most common butterflies in North America.

You can often identify them quickly by their spotted bodies and mostly black wings.

Their wings have rows of spots along the edges. Males will have mostly yellow spots, while females have more blue at the bottom.

The undersides of males and females are the same, with rows of yellow and orange.

This photo shows black swallowtail caterpillars of all different stages, or instars. Young caterpillars are mostly black with saddles, eventually becoming a striking striped and spotted beauty. These caterpillars are sometimes called the parsley worm because they use garden plants as hosts: parsley, dill fennel, carrots, celery, parsnip, and Queen Anne's lace.

This one formed a j-shape and then attached the chrysalis to our herb garden planter!

You can attract these butterflies with garden plants and a wide variety of nectaring plants like flag iris, blue mistflower, purple coneflower, and milkweed.

Papilio polyxenes
2½ - 4¼"

Eastern Tiger Swallowtail

Eastern tiger swallowtails have a striped body (like a Palamedes swallowtail) and tiger-like stripes on the upper sides of their yellow wings. Females have more iridescent blue markings on the hindwings than males, and some are much darker!

On the underside, you can see more tiger stripes, and blue and orange spots. The blue spots are the furthest line from the edges. The Eastern Tiger Swallowtail is the state butterfly of South Carolina (and several other states.)

Host plants include lilac, cottonwood, tulip, willow, sweet bay, and cherry.

Adults will nectar on joe-pye weed, milkweed, phlox, and (left) swamp azalea.

Like other swallowtails, young tiger swallowtail caterpillars mimic bird poop, developing eye spots in the final instar to confuse or mislead predators.

This butterfly also has a dark morph (differently colored version). You can distinguish the dark morph from the Palamedes because:

- It lacks the parallel body stripe and top dots
- It has a blue stripe on the innermost edge of the hindwing, while the Palamedes' inner stripe is orange.

compare to

Eastern tiger swallowtail

Palamedes swallowtail

Papilio glaucus
3⅛–5½"

Pipevine Swallowtail

Pipevine Swallowtails are dark with spotted bodies like a black swallowtails. The pipevine swallowtail has seven conspicuous orange dots under each hindwing. The upper wings have much smaller white markings on the margins, and males have an iridescent blue sheen.

These swallowtails use native pipevine (genus *aristolochia*) and arrowleaf ginger (*Hexastylis arifolia*)as host plants, and they will nectar on a wide variety of plants including the non-native pentas (*pentas lanceolata*) pictured here in red.

Caterpillars are black with raised orange/red spots. Some species of pipevine plants are toxic, and that toxicity is passed along to the caterpillars who eat the leaves. The bright orange colors on the caterpillar communicate a danger warning to potential predators, an adaptation called **aposematism**.

The chrysalis has slight winged ridges along the edges.

Battus philenor
$2^{3}/_{4}$ – $3^{3}/_{8}$”

Pieridae: Sulfurs, Yellows & Whites

Early British naturalists described these as "butter-colored flies," which likely evolved into the very word butterfly. This family includes the whites, sulfurs, orangetips and yellows.

They are easy to observe because they fly relatively close to the ground.

little yellow on camphorweed

They are regular and dependable visitors to flowers and gardens, using brassicas, legumes and mustards as host plants.

Cabbage whites, a non-native butterfly (which can be a pest on agricultural crops) are members of this family.

cloudless sulfur on morning glory

falcate orangetip

Cloudless Sulfur

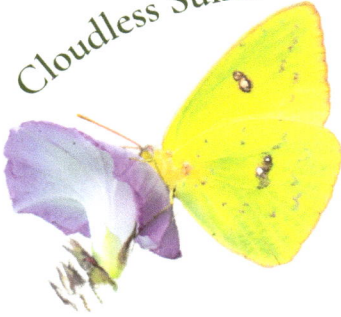

- biggest yellow butterfly in the area
- mostly yellow
- 2⅛ – 2¾"

falcate orangetip female on dewberry

Falcate Orangetip

- white above
- black spot on forewing
- male has orange tips on top wings
- female is all white
- only visible in early spring
- 1⅜ – 1½"
- both sexes have slight black margins at the edges
- from the underside they look the same.

male

Sleepy Orange

Little Yellow

- 1⅜ – 1⅞"
- slightly larger than little yellow
- rusty marks spread across the center of the wing

- 1 – 1½"
- much smaller than cloudless sulfur
- black edge
- rust spot on edge of hind wing

Falcate Orangetips are early spring butterflies that are relatively small and are almost always on the move.

Males emerge first, with bright orange spots or the curve of the forewing, and they fly low over patches of host plants.

female

male

Females are almost all white on the top, with a few small black spots. The underwings of both males and females are marbled.

Falcate Orangetip

falcate orangetip on dewberry and roundleaf bluet

falcate orangetip on black medic clover

These butterflies have only one brood a year, and the adults live for about a week, so there's a very short window to encounter them.

Falcate Orangetip on Black Medic Clover

They use members of the brassica family (cabbages, mustard, etc.) as host plants. They will nectar on whatever is blooming in early spring, like clover, bluets and dewberries.

"Falcate" refers to the slight sickle-shaped hook on the top of the forewing.

Sea Rocket

Falcate Orangetip on Dewberry

I found this one under a sea rocket plant where Falcate Orangetips were feeding. Sea rocket is an early colonizer of beach dunes and a member of the brassica family.

Anthocharis midia
$1\frac{3}{8} - 1\frac{1}{2}$"

27

The little yellow butterfly is a small sulfur with dark margins on the forewing. Their color can vary, from very pale to deeper yellow. They tend to fly right at the tops of the grass, relatively close to the ground. The butterfly below was flitting under dune plants. This butterfly looks a lot like a cloudless sulfur, but is only about half the size.

The caterpillar is green with a yellow stripe.

Little Yellow

camphorweed

frogfruit

Sometimes a group of butterflies (usually males) will gather around a patch of wet ground or shallow puddle, a behavior called "puddling." They can use their proboscis to consume minerals from the soil.

Little yellows use members of the pea family (butterfly pea, partridge pea) as host plants.

They nectar on a wide variety of plants, including dotted knotweed and frogfruit, below.

dotted knotweed

butterfly pea

frogfruit

Pyrisitia lisa
1 – 1½"

29

Cloudless Sulfur

hedge bindweed

Cloudless Sulfurs are a relatively large, almost completely yellow butterfly. They are one of the most common butterflies in the lowcountry:

seashore mallow

Adults meander north during the summer and many migrate southward in late summer or autumn. They are present in SC year-round.

flag iris

hedge bindweed

The caterpillars vary from very pale to a deep yellow, but most have a strong lateral yellow stripe and tiny dots. The wing is edged in pink. Cloudless sulfurs use plants in the pea family as host plants, as well as sennas.

cassia

blue mistflower

They feed on flowers with their long, straw-like proboscis.

hedge bindweed

Phoebis sennae
$2^1/_8 - 2^3/_4$"

Sleepy Orange

Named "sleepy" because of their lack of eye spots, these butterflies never seem sleepy! They're often moving so quickly that they are hard to capture with a camera!

snow squarestem

They vary from pale yellow to orange i appearance. In the fall and winter, you' more likely to see the deeper orange color.

It is rare to find these butterflies resting with their wings spread, but sometimes you can see the black margins at the edge of the wings when they're flying or when the sun shines through the butterfly.

firewheel

You can attract this butterfly by planting nectar plants like the asters below, or host plants in the legume family, like butterfly pea or beach pea, and even clovers. Cassias and Sennas may also attract them.

butterfly pea

This chrysalis is almost ready to eclose: you can see the black marks against the orange wings.

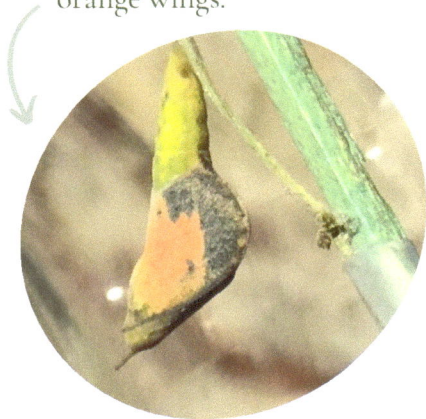

beach pea

Abaeis nicippe
$1^{3}/_{8}$ – $1^{7}/_{8}$"

Nymphalidae: Brushfoots

Also known as the Brushfoot family of butterflies, this is the biggest group of butterflies, and has several sub-families. The top side and undersides, as you can see on with this common buckeye.

The front legs of most brushfoot butterflies are very small, giving them the appearance of having four legs. The tiny front legs may be mostly sensory, helping the butterflies "taste" potential nectar sources.

Nymphalid Butterfly Families:

The Milkweed Butterflies (Danianae)

These are the butterflies that use milkweed as host plants.

Monarch

Queen

The Longwings (Heliconiiae)

These include the longwings and fritillaries.

Gulf Fritillary

Variegated Fritillary

Zebra Longwing

The Viceroys and Admirals (Limenitidinae)

Viceroy

Little Wood Satyr

The Nymphs and Satyrs (Satyrinae)

The "True Brushfoots" (Nymphalids)

American Lady

Buckeye

White Peacock

Mourning Cloak

Red Admiral

Pearl Crescent

Phaon Crescent

Painted Lady

Monarch

One of the most famous butterflies, Monarchs are in the lowcountry year-round.

While this species is famous for migration to Mexico, recent studies have shown that some of our local Monarchs overwinter instead of migrating, while others wander through on their way south.

These black-veined orange butterflies are lighter underneath, with a double row of white spots in the black border.

fall migration on groundsel

There is a wealth of information about attracting monarchs, and you can even track their migration throughout North America using a citizen science network called "Journey North."

Monarchs use milkweeds as host plants, and some milkweeds contain toxic compounds   The bright orange color serves as a warning to birds that the butterfly might be toxic.

The only native milkweed we have out on this barrier island is swallowwort, (*Cynanchum angustifolium*) pictured here.

Swallowwort

compare to

Viceroy

Monarch

The Viceroy's colors mimic the monarch, even though it carries different compounds.

To attract monarchs to your yard, plant a variety of native nectaring plants. Along the coast, Groundsel (Baccharis Hamifolia), blooms as the monarchs migrate, so it's a great addition to your butterfly friendly yard.

The chrysalis is a beautiful jade green with gold dots. Right before the butterfly ecoses, it becomes transparent

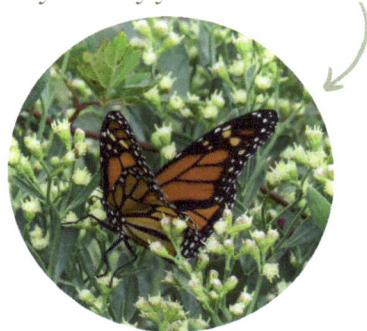

Danaus plexippus
3½ - 4"

Queen

Queen butterflies, like monarchs, use members of the milkweed family for host plants and are poisonous to some predators.

compare to

Monarch

Groundsel

Queen

Queen caterpillars are very similar to monarchs, but they have an additional, third set of tentacle-like protuberances between the head and the center of the body.

The butterfly's color is also similar to that of a monarch, but they have a little constellation of spots in the middle of the wings. They have spotted bodies.

toothache tree

toothache tree

swallowwort

frogfruit

To attract Queen Butterflies, try Swallowworts, Joe Pye Weed, Boneset, Blue mistflower, frogfruit and Toothache Trees. Groundsel (*Baccharis hamifolia*), is a great plant for such a wide variety of butterflies.

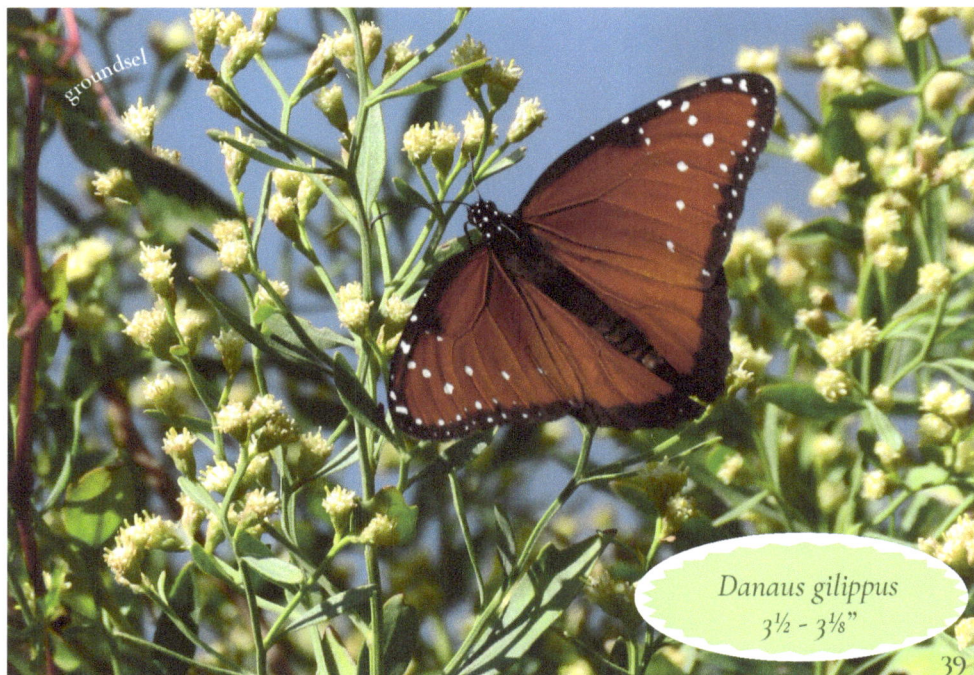
groundsel

Danaus gilippus
3½ - 3⅛"

snow squarestem

Zebra Longwing

These easily recognizable long-winged butterflies have thin yellow stripes on a dark background and narrow long wings.

There are tiny red dots on the underside, near the thorax.

They drift in a relatively relaxed manner when flying, ambling from flower to flower.

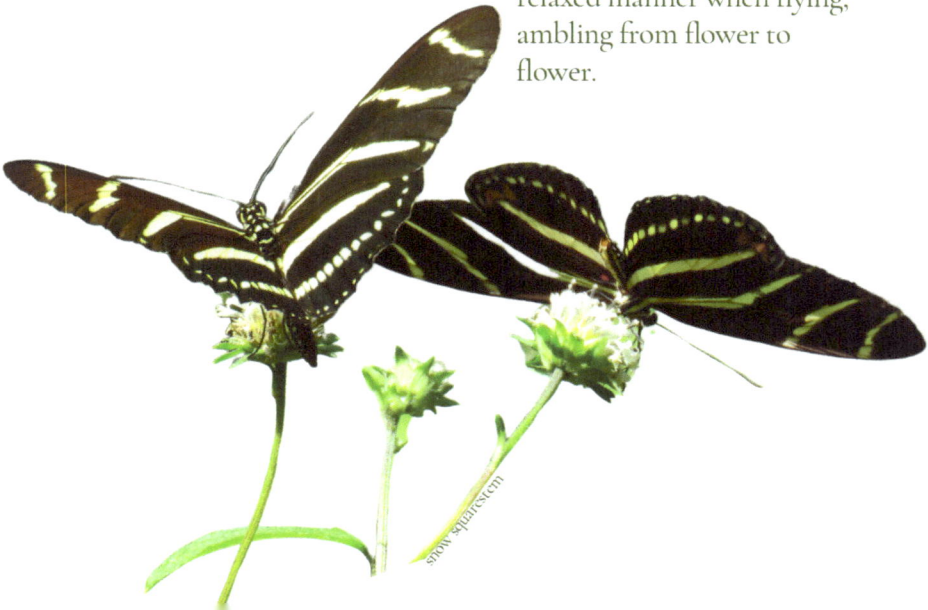

snow squarestem

Zebra Longwings are the only butterfly that eats pollen as well as nectar. This enables them to have a much longer lifespan than most butterflies.

The caterpillar is white with black branched spines.

This species likes the tropics, flittering around forests, edges, hummocks, and open areas. Sometimes they roost at night in large groups. The zebra longwing is the state butterfly of Florida.

They nectar on a variety of plants (we see them on rosemary regularly) and they use passionflower as a host plant. Also known as maypops, the plant, *passiflora incarnata*, is a favorite of gulf fritillaries too.

We've even been successful growing passionflower in containers!

passionflower

Heliconius charithonia
3½ - 3⅜"

Gulf Fritillary

Gulf fritillaries are ubiquitous along the coast in late summer and fall. If you plant passionflower (purple or yellow), you'll likely have a large crowd of these cheerful brushfoot butterflies.

Passionflower

They are orange on the top side, with some black marks. The patterned underside has brown, black, and an almost iridescent sheen.

thistle

passionflower

They lay single eggs on the leaves of the passionflower, which hatch into orange spiky caterpillars. When they are ready to pupate, they hang in a J-shape, and wriggle out of their spiny coat, leaving a pile of spines on the ground like a sweater.

Native passionflowers: passiflora incarnata or passiflora lutea, are great plants for attracting gulf fritillaries and other butterflies.

These butterflies are among the easiest to raise at home.

Purple Passionflower

Yellow Passionflower

Agraulis vanillae
2½ - 2⅞"

Viceroy

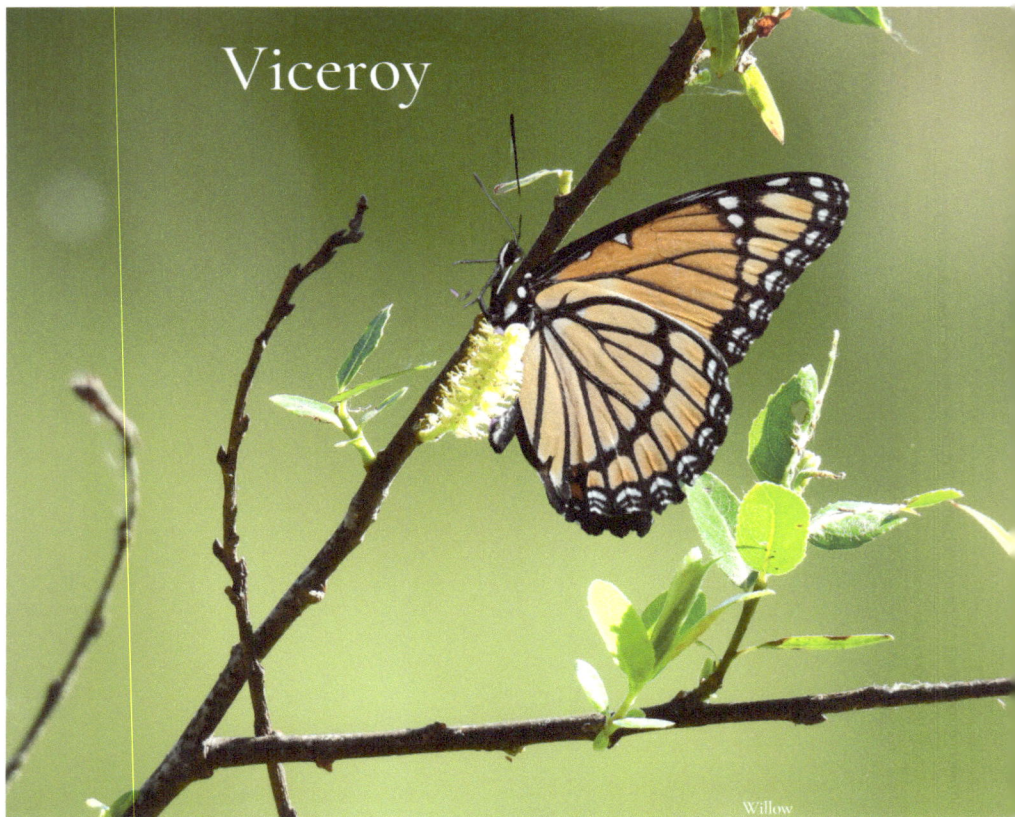

Willow

Viceroys use willows as host plants, with the caterpillars absorbing the willow's salicylic acid, making them unpalatable to birds (like Queen and Monarch butterflies.) They'll also use cottonwood and poplar trees as host plants and will nectar on a wide variety of flowers.

Wax myrtle

Viceroy caterpillars, like swallowtails, look a little like bird droppings on the leaves of host plants.

To attract viceroys, try groundsel, willow, poplar, and cottonwood trees, and nectar plants like thistle and milkweeds.

Willow

Viceroys have similar colors to Monarchs and Queens, but the Viceroy is smaller, with a black line that crosses the hindwings.

You'll find more comparisons on pages 58-59.

thistle

Limenitis archippus
2⅝ - 3"

Red Admiral

toothache tree

Like all brushfoot butterflies, the undersides of the wings (ventral sides) are quite different from the top (dorsal) sides of the wings. In this one, the ventral sides are much more muted.

The top is black/brown, and the forewing has white dots at the edge and orange stripes. The hindwings are rimmed with orange.

Red admirals often seen sunning with their wings open. They are highly territorial and will chase off other butterflies that venture into their territories. I've even had them fly right at me!

Try planting false nettle plants to attract these butterflies.

Red admirals use nettles as their host plants. When the caterpillars hatch, they disguise themselves by curling up within leaves, coming out to feed at night. During the day, the caterpillars hide by basically sewing themselves into leaves with silk.

false nettle

can you see the caterpillar in here?

When the caterpillar is ready to spin a chrysalis, it will shed the spines, dropping them to the forest floor.

The chrysalis is completely camouflaged within the nettle foliage.

Vanessa atalanta
1¾ - 3"

Common Buckeye

The Common Buckeye is anything but common in its striking beauty, detailed coloring, and conspicuous multicolored eye-spots on the back! The pale bar on the forewing is also a good field mark.

The underside of the butterfly is similar, but the hindwings are much more subdued in color.

Even the caterpillar has some of the dramatic colors of the butterfly!

Those bright eye-spots may warn off predators by implying that the butterfly is a much larger organism.

Buckeyes nectar on a wide variety of plants, and you may see them with monarchs and gulf fritillaries on groundsel during fall migration.

groundsel

We've also found them nectaring on goldenrod, tievine, climbing hempweed, yaupon, marsh fleabane, and frogfruit. Attract these butterflies with host plants of frogfruit, snapdragons, and toadflax, as well as groundsel and goldenrod for nectar.

Junonia coenia
2 - 2½"

frogfruit

American Lady Butterflies are orange on the top, with black markings along the edges. In the center of the top of the wing, there is a tiny white dot on the orange. The underside has a cobweb pattern with two large eye spots at the edge of the hindwing.

American Lady

Look how different the dorsal (top) and ventral (bottom) sides are with the wings folded up! (Same butterfly, seconds apart.)

They lay their eggs in cudweed or everlasting plants, where the larvae (caterpillars) spin weblike structures using their own silk and the flower fluff to make nests right inside the plant's flower structures. They come out of this hideout at night to feed.

The chrysalises can vary in color with golds and browns, and sometimes has tiny triangular bumps with dots and vertical lines.

Isn't it incredible that this chrysalis contains everything necessary to become an adult butterfly?!

Vanessa virginiensis
$1\frac{1}{4} - 2\frac{3}{8}$"

Painted Lady

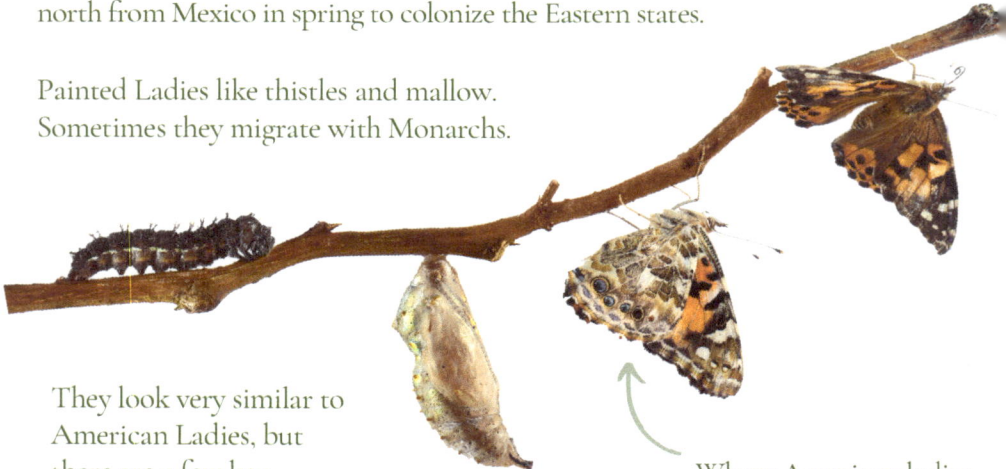

Painted Ladies are one of the most widespread butterflies around the world. They are capable of long-distance migration, migrating north from Mexico in spring to colonize the Eastern states.

Painted Ladies like thistles and mallow. Sometimes they migrate with Monarchs.

They look very similar to American Ladies, but there are a few key differences.

Where American ladies have two eyespots on the underside of the hind wing, Painted Ladies have four.

American Ladies have a tiny white dot on the orange middle section of the top of the forewing.

Painted Lady

American Lady

American Ladies have two larger eyespots

Painted Ladies do not have a white dot on orange. They have four smaller eyespots on the hindwing.

Vanessa cardui
2 - 2⅞"

53

Phaeon Crescent

This small butterfly likes moist coastal areas. The upperwings are brown with orange checkered spots and a paler orange band in the center. The underwings (ventral sides) are similar, with more checkered spots and more cream and brown.

Some phaon crescents have blue-green marks on their thorax.

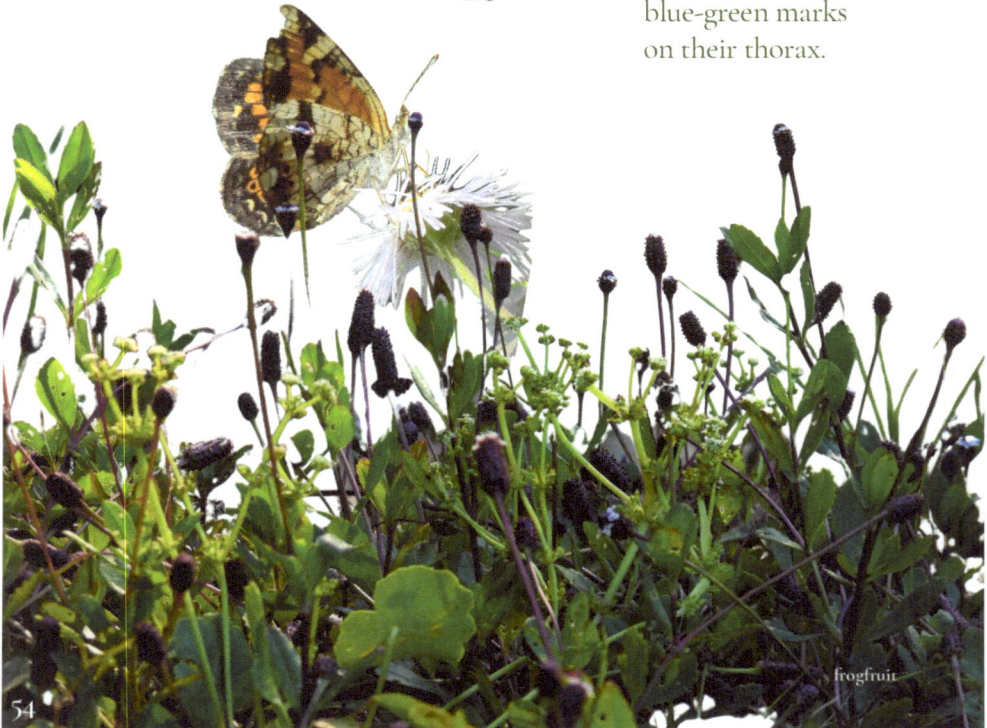

frogfruit

Males patrol
open areas
over host
plants
waiting for
females.

compare to Pearl Crescent

Pearl crescents are similar,
but lack the pale marks on
the forewings.

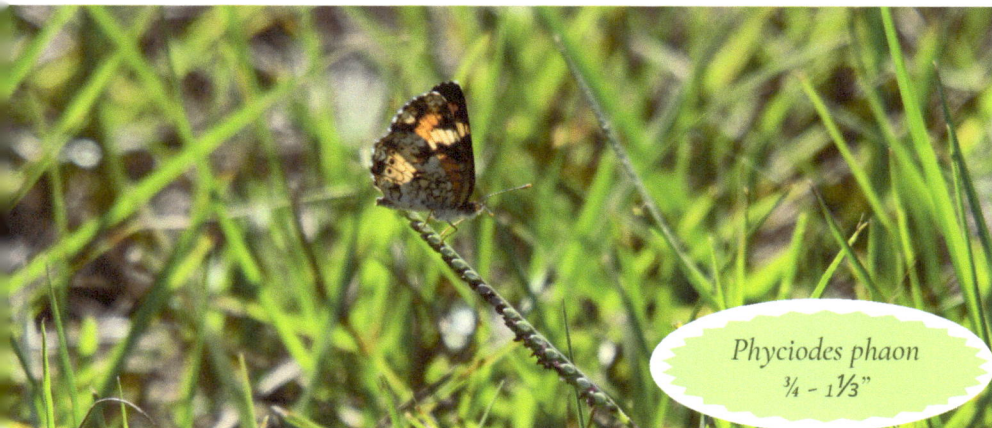

You can attract phaon
crescents with frogfruits,
which they use almost
exclusively as host plant.
These tiny plants attract a
wide variety of butterflies
and skippers and make a
wonderful native
groundcover more diverse
than a traditional lawn.

Phyciodes phaon
¾ – 1⅓"

Little Wood Satyr

Like its name suggests, this little butterfly is most often found at the edges of forests.

They bask with open wings, and each part of the wing has a pair of eyespots circled with yellow.

Caterpillars have tiny white horns, dark side stripes, and a dark line down the back. Tiny hairs protrude from the bumps .

They have a skipping, bouncy flight near the ground at the edges of the forest, which is why they were probably named for the Greek nature sprite of the forest.

The butterfly in the top right photo landed right on my arm while I was photographing. The one in the middle was waiting for a mate suspended on the grass.

They are usually inconspicuous as they flit along the grasses and leaf litter.

Megisto cymela
$1\frac{1}{2}$ - $1\frac{7}{8}$"

Nymphalid Comparisons

Monarch

- White dots along edges and head
- Large orange sections with stained glass effect on top
- Hindwing with unbroken top-bottom sections
- Male has tiny spots on hindwing
- Nectars on goldenrod, camphorweed, groundsel

- Slightly darker shade of orange than monarch
- White spots on edges and scattered all over top wing
- Black lines very faint: no stained glass effect
- Smaller orange sections on top
- Slightly smaller than monarch
- Nectars on clover, goldenrod

Queen

- White spots on edges and in a row on top wing and smaller orange sections on top
- Smaller than monarch
- Nectars on willow and other fall plants
- Hind wing has a line bisecting the lines from the thorax

Gulf Fritillary

- Top wing mostly orange
- Underside has bright metallic spots
- Thorax is orange
- Only a few white spots on wing, all near head
- Nectars on frogfruits, groundsel, camphorweed, goldenrod

Viceroy

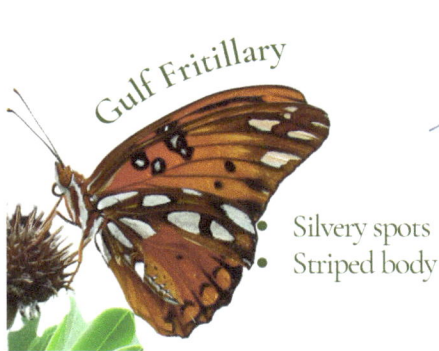

Gulf Fritillary

- Silvery spots
- Striped body

Monarch

- Stained glass
- Heavy bars on top
- No median line

Viceroy

- Median line bisecting hind wing

Queen

- Fainter upper lines on top wing
- More white dots on wing

Painted Lady

- Upper wing orange and brown
- Four eyespots on hind wing

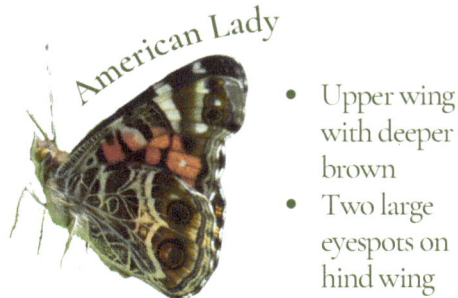

American Lady

- Upper wing with deeper brown
- Two large eyespots on hind wing

Buckeye

- Paler overall
- One large eyespot on upper wing

Red Admiral

- Mostly brown
- Red spot on upper wing with white spot

Lycaenidae: Gossamer Wings

The Lycaenids are also known as the Gossamer Winged Butterflies: The Blues, Hairstreaks, and Coppers.

Most rest with their wings folded, occasionally rubbing the hindwings back and forth.

They have the greatest defensive strategy: a decoy second head at their back end!!

There are usually eye spots near the tail filaments. The spots and filaments mirror the head, so they look like they have two heads!

Many have tail filaments

Blues

Spring Azure

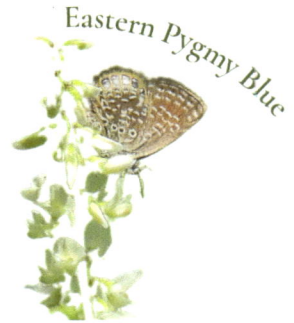

Eastern Pygmy Blue

Hairstreaks

Many Lycaenids also have striped antennae.

blue mistflower

White M hairstreak

blue mistflower

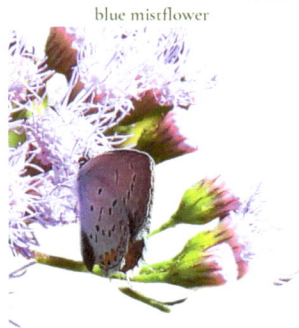

The confusing eye-spots at their back end could fool predators like these anoles into taking a bite from a wing instead of the head!

The mistflower was covered with butterflies, which provided great cover for some hungry anoles.

The "head fake" strategy worked for the White M hairstreak below: see how the back end (with the fake eye-spots and antennae) has been bitten off, but the butterfly survived!

Carolina anole in blue mistflower

white sweetclover, *Melilotus alba*

Eastern Pygmy Blue

This tiny butterfly is most often found in and around the salt marsh.

The top is a coppery brown, and the underside of the hind wing has four silvery black spots at the edge.

These butterflies are seriously tiny: the flowerheads on the erigeron below are only about a centimeter wide. It's the smallest butterfly in the east.

oakleaf fleabane

salicornia

Eastern pygmy blues are habitat specialists, so if you're lucky to live along the marsh, you might spot it along the salicornia (saltwort) which is its host plant.

And how cool is this incredible adaptation? Eastern Pygmy Blue caterpillars can withstand salt water inundation, so they can tolerate an occasional tidal washover!

I keep looking for the caterpillars, because according to the University of Florida, they are tended by ants, who protect them in exchange for the honeydew fluid they secrete. I'd really like to see that!

The caterpillars are light green with tiny white bits that blend perfectly with the salicornia.

Sea oxeye daisy and oakleaf erigeron are great nectar plants.

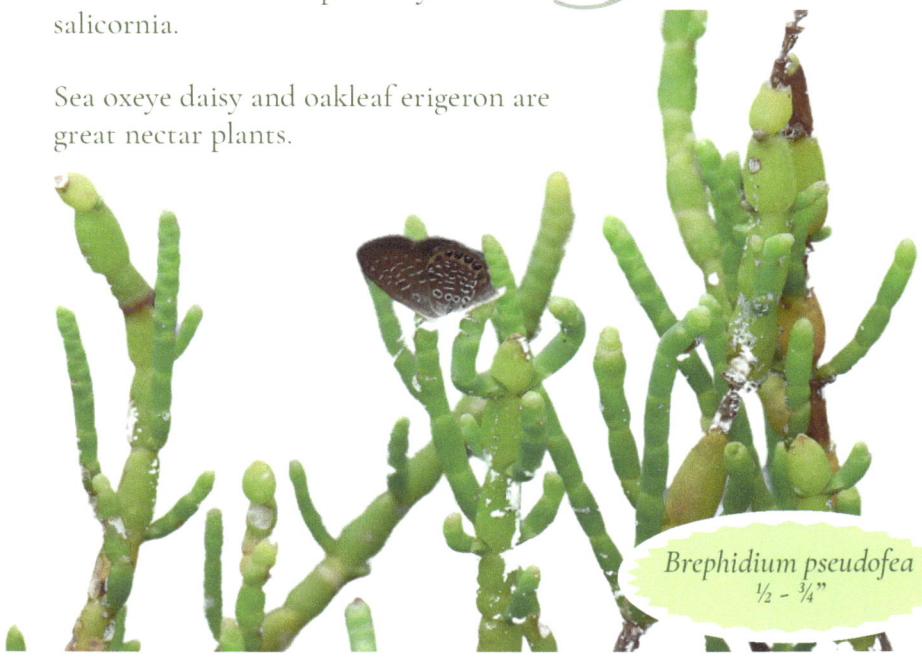

Brephidium pseudofea
½ - ¾"

63

White M Hairstreak

We have several species of hairstreaks here in the Lowcountry, and I'll confess they befuddle me most of the time: they fly quickly and erratically and they look similar.

White Sweet Clover

Our white clover is covered with hairstreaks at certain points of the year, and one is the White M hairstreak. See how the tail marks sort of make an M shape?

It's a shame that we rarely see this butterfly with open wings: just look at the incredibly vivid blue on the upper side of those wings! It's hard to believe they belong to the same butterfly! They flit so quickly between the flowers that you hardly even get a glimpse of that vibrant blue: it's like a secret they keep all for themselves.

Parrhasius m album
1⅛ - 1½"

Gray Hairstreaks
lack the blue on the
hindwing and are
one of the only
hairstreaks that bask
with open wings.

marsh fleabane

Gray Hairstreak

sea oxeye daisy

Strymon melinus
1 – 1¼"

Hesperiidae: Skippers & Duskywings

The butterflies in the family Hesperiidae are called skippers. Their quick skipping flight habits give them their name. If you get the chance to sit in front of a field of blooms being visited by skippers, you're in for a treat as they flit from flower to flower!

Skippers and Duskywings

Long-tailed skipper

blue mistflower

Fiery skipper

Silver spotted skipper

They have solid muscular bodies.

The skippers have hook-tipped antennae, where most butterflies have clubbed antennae.

Ocola skipper on groundsel

Some rest with wings closed, others with wings open.

It's hard to find caterpillars in this family because they feed most often at night. The greenish caterpillars with oversized heads curl up within leaves during the day.

Long-tailed skipper caterpillar in rolled leaf, outside and inside

This photo looks up the end of a rolled leaf. The black is the head of the caterpillar, and on the left you can just see the silk threads that the caterpillar uses to cement itself into the leaf.

Checkered skipper

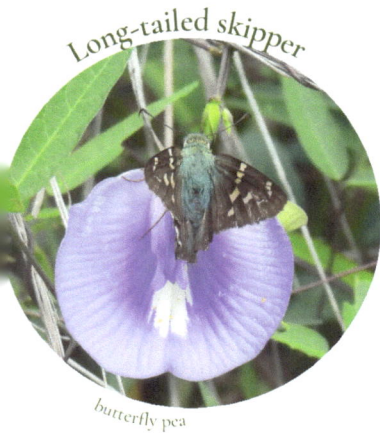

Long-tailed skipper

butterfly pea

Duskywings rest with open wings.

yaupon

goldenrod

Long-tailed Skipper

Long-tailed skippers are the most easily identifiable skipper with their long tails and iridescent green bodies.

They use members of the pea family for host plants.

There are pale squarish dots on the upper forewings.

Caterpillars can spit a bitter fluid to discourage predators!

goldenrod

Urbanus proteus
1½ - 2"

68

Checkered Skipper

There are several sub-species of checkered skippers, but they all have checkered edges. They feed and rest during the day with open wings. At night, the wings will be tightly closed. Many have blueish gray fuzzy bodies.

Pyrgus communis
¾ - 1¼"

Fiery skipper males have toothed brush-stroke margins at the edges of the hindwings. The underside has small dots.

Fiery Skipper

Skippers can really be confusing, There are a bunch of grass skippers with brown or orange and they often rest with wings partly open: sort of a triangular jet-plane stance. Definitive identification of these usually involves poring over field guides, looking at their wing margins and ranges.

The fiery skipper is bright orange with some small spots on the wing.

Hylephila phyleus
1 - 1½"

Silver Spotted Skipper

The silver spotted skipper is a large skipper, easily recognizable when the wings are folded by the large white spot on the underwing. The same spot appears orange on the top side. They use members of the pea family like ticktrefoil and downy milkpea as host plants.

Epargyreus clarus
1½ - 2⅝"

Salt marsh skippers have a distinctive
light bar in the center of the wing.

Panoquina panoquin
1¼"

Salt Marsh Skipper

Ocola Skipper

Ocola skippers have long
narrow forewings that
extend way past the body.

Panoquina ocola
1⅜–1½"

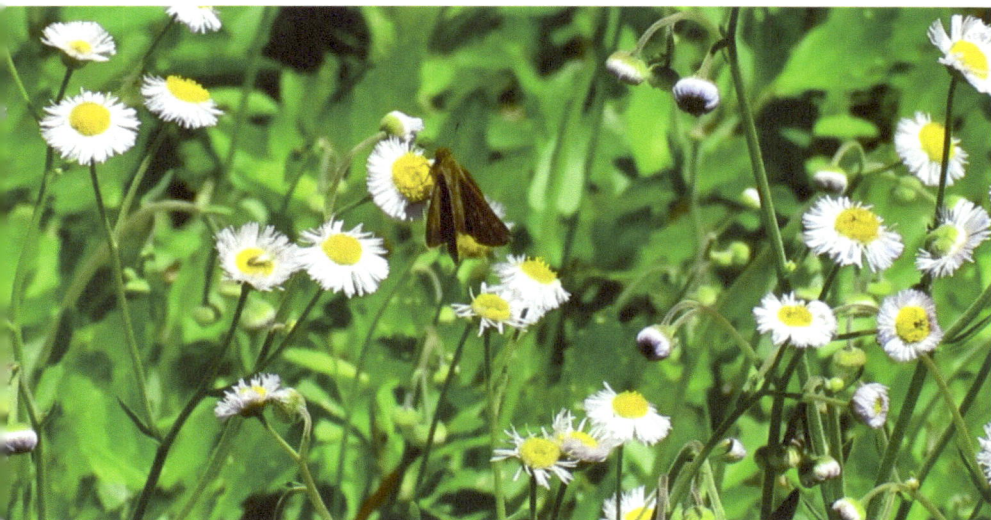

Eufala Skipper

Eufala skippers have 3-5 creamy dots on their wings.

There are a LOT more skippers, and to be frank, I have a hard time identifying these, even with field guides or identification apps!

Don't let your inability to definitively identify them hamper your enjoyment of watching them skip from flower to flower!

Lerodea eufala
1 - 1¼"

Duskywings

The same challenges apply to the duskywings, a group of spread-winged skippers that can be hard to tell apart.

They rest with open wings, and are similarly shaped.

Horaces Duskywing

They have a mothlike appearance, but are active during the day.

In our area, you are most likely to see Zarucco, Horace's, or Wild Indigo Duskywings.

Erynnis horatius
1⅜ - 1⅞"

74

Moths

Moths are also Lepidoptera, and any lepidopterans that don't qualify as butterflies are classified as moths.

Moths tend to have feathery antennae and fatter and fuzzier bodies. Moths are important night-time pollinators, and their fuzzy bodies may make them even more effective at gathering pollen than their butterfly counterparts! There are far more moths than butterflies, and we are just beginning to understand all the roles they play in the ecosystem!

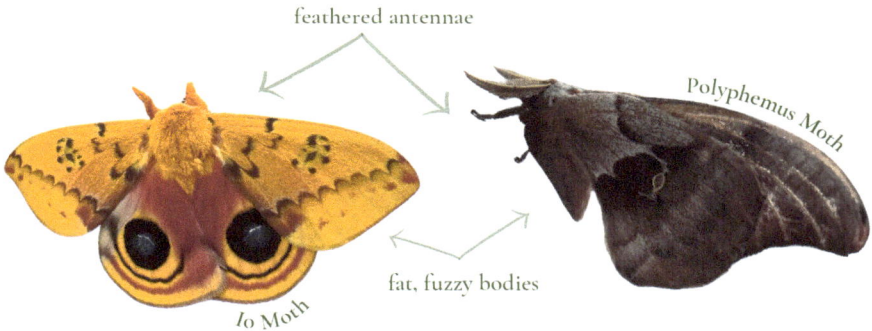

feathered antennae

Polyphemus Moth

fat, fuzzy bodies

Io Moth

Nocturnal predators call for different defense strategies. Some moths have eye spots that they can "blink" by closing or opening their forewings over the spots.

Some moths have developed a sense of hearing so they can listen to (and evade) bat sonar.

This Eastern Screech owl has captured a polyphemus moth and brought it home to her nest.

Giant Silkworm Moths

The giant silkworm moths (Luna, Polyphemus, Io, Imperial) are all huge, with wingspans that could be as large as 7 inches across! These moths do all their eating as caterpillars (the adults don't even have mouth parts!) Most of them are only in the adult stage for a very short time, about a week or so.

Luna Moth

Luna moths have long tails, which may actually scramble bat sonar or confuse bats the way the hairstreak's tail filaments do. The larvae is bright green, and this moth has just eclosed and is waiting for its wings to fill.

Sometimes we only see these beauties after they have perished, like the top luna moth, found on the forest floor of Audubon's Beidler Forest.

Actias luna
3 - 4.5"

Polyphemus Moth

Polyphemus moths are also only in the adult stage for about a week, but their fascinating larvae create an incredible cocoon out of silk and leaves.

Antheraea polyphemus
3¾ - 5¼"

Imperial moth

Imperial moths look a little like fallen poplar leaves, which is one of the trees they use as hosts.

Their voracious caterpillars feed on a variety of plants, like this one on a pine branch. Be careful with the caterpillars: some people have a painful reaction to the tiny hairs!

Eacles imperialis
3 - 4¼"

Io moth

Like a Polyphemus moth, Io moths have large eyespots that can be covered when the wings are closed so the moth can flash the "eyes" at the right moments.

Io moths have striking brushy caterpillars. After feeding on leaves, they drop to the forest floor and shed their bristles when they pupate.

Automeris io
2 - 3⅛"

Tussock Moths

Orgyia sp
¾ - 1¼"

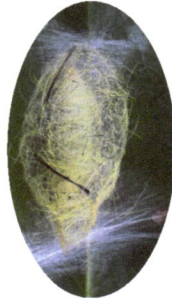

Tussock Moths leave furry cocoons on homes and in garages that can be hard to remove. The caterpillar hairs are capable of stinging, so don't pick them up!

Sphinx Moths

There are many species of sphinx moths, or hawk moths. These have an incredible ability to hover, and some moths in this group are referred to as hummingbird moths, because they are large enough to look like hummingbirds when they're feeding, and they have a long proboscis!

Pandorus Sphinx

Eumorpha pandorus
3 ¼ - 4 ½"

Sometimes these moths startle people by zooming around them right at dusk.

Carolina Sphinx

The Carolina Sphinx moth is known as a tobacco hornworm in its larval stage. The caterpillar has a large horn and it feeds on tobacco, tomatoes, and food crops.

Manduca sexta
3 ¾ - 4¾"

Pink-stripe Oakworm Moth

These fly during the day and use oaks as their host plant, laying their eggs under the leaves. Like the giant silkworm moths, adults don't feed at all. They pupate in the ground over the winter.

Anisota virgiensis
1½ - 2½"

Forest Tent Caterpillar

Forest tent caterpillars, despite their name, do not make silk tents. Instead they create a silk sheet where they can all stay together. They follow each other using scented silk threads.

As they progress through later instars, they become progressively more independent.

Malacosoma disstria
1 - 1¾"

Ailanthus Webworm Moth

Ailanthus Webworm Moths have an incredible adaptation: when flying, they look like wasps. Their larvae live in communal tent-like webs.

Atteva aurea
¾"

Butterfly and Moth Friendly Yards

There is a lot you can do to help butterflies and other insects right in your own backyard.

Reduce or Eliminate Pesticide Use

Most pesticides are toxic to a broad spectrum of insects, including all life stages of butterflies. Whether it's sprayed from a misting system or from a portable backpack, a broad-spectrum pesticide will kill butterflies, dragonflies, frogs, snails, and more. Try reducing mosquitoes by eliminating stagnant water, planting repellent plants, or using organic alternatives. The Xerces Society has lots of suggestions!

Reduce the Amount of Turf grass to Create Natural Areas

If you haven't checked out the ideas and website of the Homegrown National Park movement, it's worth a look. Grass lawns are labor intensive and consume lots of resources (water, fuel, etc.) to become a monoculture. Swap some (or all) of your lawn for some native plants and trees that are host plants or nectar sources for butterflies and bees, and try making your grass lawn a little smaller each year.

Steer Cleer of Herbicides

Herbicides, particularly those containing glyphosate, are also a danger to bees and other pollinators, including butterflies. Swallowtail eggs near plants treated with glyphosate are significantly less likely to hatch. Pull weeds by hand, allow some "weeds" to flourish for the benefit of pollinators, or use physical barriers like raised beds to reduce unwanted plants.

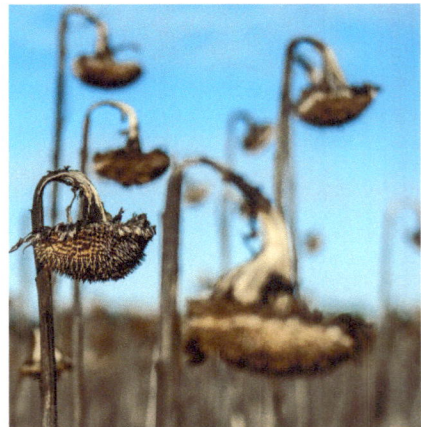

Plant a Variety of Native Plants and Trees

The more biodiverse your plant life is, the more diverse your insect population will be. Your local native plant nursery likely knows what the best bang for your buck is, and can help you decide what plants are best for your growing conditions. Trees provide both shade and habitat. Oaks are a great choice, as are sweet gum, hackberry, willow, redbud, persimmon, walnut, hickory, poplar, and maple.

We suggest some great host plants on pages 84-85.

Live Oak

Sugar Maple

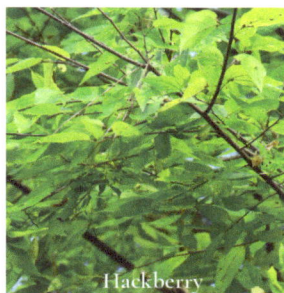
Hackberry

Watch Out For Neonicitinoids

Neonicitinoids, aka neonics, are systemic pesticides used to treat plants and seeds. They are often used preventatively by the agriculture industry, and they are some of the most toxic insecticides ever used. If a plant was treated with neonics (even as a seed), and you plant it, you can inadvertently poison the very butterflies you're trying to attract. Buy plants and seeds from reputable sources and always check to see that they have not been treated with neonics.

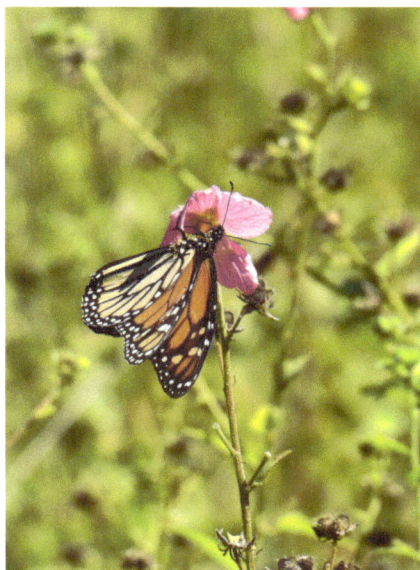

Install Native Plants for Nectaring

Go beyond host plants, and provide multiple sources of nectar. Native plants are best because the plants and the butterflies have evolved together in the same ecosystem in amazing and unique ways. Nectaring plants often have tubular flowers shaped perfectly for the proboscis of specific butterflies. Select plants that bloom in a variety of colors, and try to have something blooming all year.

Leave the Leaves and Mow Less Often

Many butterflies and moths pupate right in the leaf litter. Other insects spend the winter in the leaf litter as well, and it's good mulch for your garden plants. If you do need to rake your leaves, try leaving a "pollinator pile" where insects and larvae can hatch safely out of the way of lawnmowers. You might even find more fireflies!

Let the Plants Stay Sloppy Over the Winter

Dead leaves may be hiding chrysalises tucked in the branches. Some of these will overwinter, or eclose late in the season: leaving the dead stalks will allow these to stay off the ground, and stalks provide shelter for other insects in winter.

Add Fruit, Mud, and Water

A shallow dish of sand or mud will attract butterflies for puddling, and you may be able to "sweeten the pot" with some sliced fruit. Butterflies can't land on water to drink, so wet pebbles can provide landing areas in a deeper birdbath. If you do provide water, and not just wet sand, be sure to empty every two days or so to keep mosquito larvae from developing.

Reduce the impact of artificial light

Artificial light can disorient moths and disrupt their mating cycles. Try to keep your outdoor lighting minimal. Try to reduce the amount of light that escapes from your house by using window coverings.

Think about installing light bulbs with warm tones, which are less visible to wildlife. (We did this and the number of spiderwebs around our house significantly decreased.) Downward-facing lights can also reduce your light pollution. Installing timers can ensure your lights are off. There are lots of ideas in your local chapter of Starry Skies South.

Some favorite host plants

Swallowwort

Swallowwort, *Cynanchum angustifolium*, is the only native milkweed on the barrier island where I live. There are many other milkweeds you can plant to attract monarchs, including swamp milkweed, but be sure it hasn't been treated with neonicitinoids.

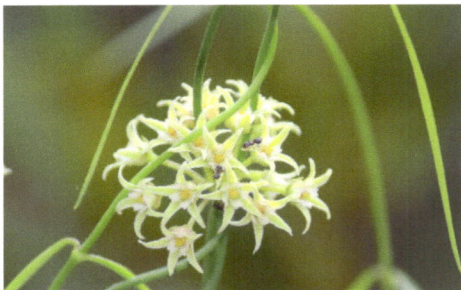

Toothache Tree

Also known as Southern Prickly Ash or Hercules Club, *Zanthoxylum clava-herculis* is a host for giant swallowtails and an incredible nectar plant for a huge variety of butterflies. Eight different species in this book are photographed on this tree!

False Nettle

False nettle, *Boehmeria cylindrica*, is a great host plant for red admirals, eastern commas, and others.

Passionflower

Both yellow passionflower, *Passiflora lutea*, and purple passionflower (maypops), *Passiflora incarnata*, are hosts to fritillaries and zebra longwings.

Groundsel

Groundsel, aka salt myrtle, *Baccharis Hamifolia*, is an incredible attractor of all sorts of butterflies, especially during migration. There are both male and female plants. The white seeds are the female flowers; the male flowers have yellow blooms and appear on separate shrubs.

Frogfruit

Frogfruit, *Phyla nodifera*, is a host plant to phaeon crescents and several skippers. Many other insects use them as nectar plants. It's a great groundcover!

Blue Mistflower

The patch of blue mistflower, *Conoclinium coelestinum*, in my yard is usually covered with butterflies in late fall.

Seaside Goldenrod

Another fall butterfly nectar magnet, seaside goldenrod , *Solidago sempervirens*, is always covered with butterflies.

Great Web Resources

iNaturalist, Seek, and iNaturalist.org

Download the Seek app on your phone; you can use your camera to identify butterflies and caterpillars you're not sure of. When you create an iNaturalist account, fellow observers can help you check your observations. I can't overstate the value of this resource!

The Xerces Society for Invertebrate Conservation

There is a wealth of information at Xerces.org! They have lists of native plants that support butterflies and other invertebrates, suggestions for alternatives to pesticides, webinars, and more.

Homegrown National Park

At homegrownnationalpark.org, you can learn about ways to make your yard (or patio, or balcony) into a habitat that supports biodiversity.

The grassroots collective aims to get backyard "parkland" registered with a certain number of acres per year. You'll find publications by founder Doug Tallamy on the book list on page 88. This is just an incredible resource for everyone.

Starry Skies South

Starry Skies South, and DarkSky.org offer great information about light pollution and its hazards to humans and wildlife, with wonderful suggestions for limiting your light footprint. You can even certify your yard!

South Carolina Wildlife Federation

The South Carolina Wildlife Federation has a wealth of information about building a backyard habitat, and all sorts of incredible classes. One of their classes (at the South Carolina Botanical Garden provided me with a lot of these photos. Join, follow, donate- they're great!

HGIC Clemson

The Home and Garden Information Center at Clemson University is a great resource for learning all about butterflies and native plants! Once you start exploring their resources, it's hard to stop. Clemson Ext. also has great classes, including the SC Master Naturalist Program.

The North American Butterfly Association

The North American Butterfly Association is a non-profit dedicated to the conservation of wild butterflies and their habitats.

Journey North

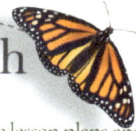

Journey North is another citizen science database that provides a chance for you to both enter data and track monarch migration in almost real time. They also provide lesson plans and other resources. www.journeynorth.org

Audubon South Carolina

Check out Audubon South Carolina's bird friendly gardening suggestions: plants that are good for birds are often great for butterflies and caterpillars. Their South Carolina centers at Beidler and Silver Bluff are worth a visit. Several butterflies in this book were photographed at Beidler! Join, follow, visit, donate!

Alabama Butterfly Atlas (and Native Plant Atlas)

The Alabama Butterfly Atlas provides lots of information about each species, and an incredible database of butterflies and host plants. If you are looking for more information about a particular species, check them out.

Butterflies and Moths of North America

This is a more technical citizen-science program for entering your butterflies, sharing photos, and keeping lists, but if you like listing and learning more about each species, this is a great option. butterfliesandmoths.org

Entomology and Nematology Department at the University of Florida

This is also an incredible resource for information about life stages of different butterflies. They had the ONLY example of a photo of a falcate orangetip caterpillar I was able to find on the web, and they have great species accounts.

Wildflower.org: The Ladybird Johnson Wildflower Center

The Ladybird Johnson Wildflower Center at the University of Texas at Austin is a great place to learn about native plants that support wildlife and pollinators!

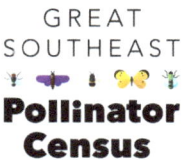

The Great Southeast Pollinator Census

This is another community science project gathering data about pollinators and native plants. Participants can enter data and learn about creating sustainable pollinator habitat, and they have incredible identification resources. Join and participate!

Nature Walks with Judy

Our website has lots of videos that show butterflies at all stages of life, pupating and eclosing and caterpillars chowing down and more. All the links to these resources are right there too. Use your phone's camera app to scan the code here.

Books Worth Reading

Nature's Best Hope: A New Approach to Conservation That Starts in Your Yard, by Doug Tallamy

Need some inspiration for making your yard butterfly-friendly? This book is just the beginning; anything by this author will inform and inspire you.

Raising Butterflies and Moths in the Garden, by Brenda Dziedzic

This is a comprehensive look at the life cycles and ways to attract butterflies to your garden. Once you've identified your butterflies, this book will help you understand exactly what the caterpillars need to live their entire life cycle in your yard.

Garden Insects of South Carolina

Written by a local team from the Lowcountry Biodiversity Institute, this book looks at beneficial species and potential pests.

Butterflies of North America (Kauffman Guide)

A great field guide for identifying butterflies in your yard.

Butterflies through Binoculars: a Field Guide to Butterflies of the East, by Jeffrey Glassberg

Another great field guide, with clear range maps and side-by-side photos to aid in identification.

The Language of Butterflies: How Thieves, Hoarders, Scientists and Other Obsessives Unlocked the Secrets of the Worlds Favorite Insects, by Wendy Williams

This is a great, fascinating dive into the world of lepidopterans and the history of the study of insects and butterflies.

A Butterfly Called Hope

A childrens book by NYT Bestselling author Mary Alice Monroe with a sweet story of a girl raising a monarch and a lot of info about monarchs and migration.

How Insects Work: An Illustrated Guide to the Wonders of Form and Function, by Marianne Taylor

This graphic and photo-rich book is an explanation of the anatomy of insects and how different organs function, with a lot about metamorphosis.

The Nature of Oaks, by Doug Tallamy

The author takes a seasonal look at the benefits oak trees provide the landscape and a strong argument for planting them to promote biodiversity.

The Girl Who Drew Butterflies: How Maria Merion's Art Changed Science, by Joyce Sidman

200 years before Darwin, at a time when women interested in caterpillars might be burned at the stake as witches, a young artist named Maria Sybilla Merian scientifically observed the metamorphoses of insects, upending long-held Aristotelian beliefs (like spontaneous generation of insects!) This book is beautifully illustrated with Merian's art.

Caterpillars of Eastern North America, by David Wagner

A comprehensive field guide to caterpillars, with good info about identification, photography, and care of caterpillars.

There are many more good resources listed on pages 96 and 97.

Photo credits

I am so grateful for for all of the iNaturalist photographers who were incredibly gracious about sharing their work and their knowledge. The iNaturalist community of photographers has entered enough data that the AI works pretty well to verify the identification of my own photos and allows me to check anything I can use from Canva. Except for the following, the photos are mine, taken on Dewees Island, at Audubon SC's Beidler Forest, or at the SC Botanical Garden at Clemson University.

2	Author bio-pic	© Reggie Fairchild
6	Black Swallowtail Instars	© Dori Wagner Eldridge
8	Black Swallowtail/Assassin Bug	© Dr. Merle Shepherd (background removed)
9	Viceroy Butterfly	© Alialura from Getty Images via Canva.com (background removed)
9	Palamedes Swallowtail caterpillar	© Dr. Merle Shepherd (background removed)
9	Tiger Swallowtail caterpillar	© Michael Harrison via iNaturalist with permission, observation 186695345 (background removed)
10	Eastern Tiger Swallowtail	© Artiste9999 from Getty Images via Canva.com (background removed)
10	Viceroy	© Photosbyjimn from Getty Images via Canva.com (background removed)
10	Common Checkered Skipper	© Ken Carman via iNaturalist with permission, observation 59470177 (background removed)
10	Zebra Longwing	© Patty_C from Getty Images via Canva.com (background removed)
11	White Peacock	© Brian Magnier from Getty Images via Canva.com (background removed)
11	Eastern Tiger Swallowtail	© Peter Weiler from Pexels via Canva.com (background removed)
11	Viceroy	© Ca2Hill from Getty Images via Canva.com (background removed)
12	Swallowtail cat. w/ osmeterium	© Prettyzhighi from Getty Images via Canva.com (background removed)
15	Giant Swallowtail cat. w/ osmeterium	© Krystof Zyskowski via iNaturalist with permission, observation 6873508 (background removed)
19	Black Swallowtail Instars	© Dori Wagner Eldridge
19	Black Swallowtail caterpillar	© Jason Ondreicka from Getty Images via Canva.com, (background removed)
21	Small tiger swallowtail caterpillar	© Jason Ondreicka from Getty Images via Canva.com
21	Large swallowtail caterpillar	© Michael Harrison via iNaturalist with permission, observation 186695345
21	Eastern Tiger Swallowtail	© Steven G via inaturalist with permission, observation 205190384 (background removed)
21	Eastern Tiger Swallowtail Dark Morph	© By Randy Emmit via iNaturalist with permission, observation 205280929 (background removed)
24	Cabbage white	© Brad Cox from Getty images via Canva.com
28	Little Yellow caterpillar	© Dan Northcut via iNaturalist with permission, observation 204205524
28	Little Yellows puddling	© Strix_v via iNaturalist with permission, observation 30350186
33	Sleepy Orange Chrysalis	© Ellen Reeder via iNaturalist, with permission, observation 181608089
35	Viceroy	© Alialura from Getty Images via Canva.com (reversed and background removed)
35	White Peacock	© Brian Magnier from Getty Images via Canva.com (background removed)
37	Monarch Chrysalis green	© skhoward from Getty Images Signature via Canva.com (background removed)
37	Monarch Chrysalis orange	© Leena Robinson from Karolina Images via Canva.com

38	Queen caterpillar on Milkweed	© Marleigh Fletcher via iNaturalist, with permission, observation 186492133
38	Queen caterpillar	© Weber from Getty Images via Canva.com (background removed)
40	Zebra Longwing (side view)	© Patty_C from Getty Images via Canva.com (background removed)
41	Zebra Longwing caterpillar	© Jan Johnson via iNaturalist with permission, observation 167081356
44	Viceroy (side view)	© photosbyjimn from Getty Images via Canva.com, background removed and replaced with Willow photograph by author
45	Viceroy caterpillar	© Royal Tyler via iNaturalist with permission, observation 16165833
48	Common Buckeye caterpillar	© Judy Darby from Getty Images via Canva.com, background removed
49	Common Buckeye caterpillar	© Sky F from Getty Images via Canva.com background removed
52	Painted Lady Life Cycle	© Science Photo Library via Canva.com
56	Little Wood Satyr caterpillar	© Nikolette Toth via iNaturalist with permission, observation 75958041 (background removed
58	Viceroy	© Alialura from Getty Images via Canva.com (reversed and background removed)
58	Viceroy side view	© photosbyjimn from Getty Images via Canva.com, background removed and replaced with Willow photograph by author
60	Spring Azure	© WildLivingArts from Getty Images Signature via Canva.com
63	Eastern Pygmy Blue caterpillar	© JF Butler University of Florida IFAS, used per their web guidelines for educational purposes
64	White M. Hairstreak blue side	© Thomas Boyd via iNaturalist with permission, observation 203310350 (background removed)
65	Gray Hairstreak with open wings	© Jack Cochran via iNaturalist with permission, observation 205478509 (background removed)
68	Long tailed skipper caterpillar on leaf	© Celeste Ray via iNaturalist with permission, observation 33310600
69	Checkered Skipper on flower (center photo)	© Ken Carman via iNaturalist with permission, observation 59470177, background removed
69	Checkered skipper on leaf	© debibishop from Getty Images Signature via Canva.com
76	Luna Moth	© Jack Cochran via iNaturalist with permission, observation 38420747 (background removed)
76	Luna Moth eclosing	© Brit Burt, with permission (background removed)
76	Luna Moth caterpillar	© Holcy from Getty Images via Canva.com
76	Polyphemus caterpillar	© Michael S. Price via iNaturalist with permission, observation 206263184 (background removed)
77	Io Moths with closed wings	© Nick Tobler via iNaturalist with permission, observation 203967791 (background removed)
77	Io Moth pupating	© Stephanie Cantle, with permission
77	Io Moth caterpillar on holly	© Stephanie Cantle, with permission
78	Pandorus Sphinx	© Rebecca Young Narkiewicz, background removed
78	Carolina Sphinx Moth	© Michael Harrison via iNaturalist, with permission, observation 3969439, background removed
79	Pink Striped Oakworm Moth Caterpillars (both)	© Dan Northcut, via iNaturalist, with permission, observation 172347523
79	Forest Tent caterpillars in large mass	© Jack Cochran, vi iNaturalist, with permission, observation 110998404

Glossary

abdomen	the lower third of the butterfly body; contains digestive tract and reproductive organ
adaptation	a technique an organism uses to make the most of its environment or improves its ability to survive
anole	a lizard native to the southeastern US
antennae	appendages on a butterfly's head used to detect chemical signals, smells, and wind
aposematism	bright colors on an organism advertise that it is toxic or noxious
biodiversity	the variety of living organisms in a given area, ecosystem, or habitat
bisect	to divide in half
brassica	a plant; a member of the cabbage/mustard family
camouflage	an organism's ability to blend it with its surroundings
chemical receptors	a specialized sensory nerve to detect chemical signals
chrysalis	the pupal stage of a butterfly (in moths, it's called a cocoon)
clasper	specialized claw-like appendages of caterpillars to use for holding leaves and climbing
cocoon	the pupal stage of a moth (in butterflies, it's called a chrysalis)
compound eyes	eyes that are made up of lots of individual lenses; it enables insects to detect motion from a wide range
eclose	when a butterfly or moth emerges from from a chrysalis or cocoon
evert	when a caterpillar sticks out its osmetarium
exoskeleton	the external shell of an invertebrate organism
eyespot	large eye-like spots on the wings of a butterfly
falcate	hooked or sickle-shaped
filaments	small hairs that protrude from the end of a butterfly's wing
forewing	the front wings of a butterfly (closest to the head)
glyphosate	a widely used herbicide that can harm invertebrates and wildlife
herbicide	a substance used to intentionally kill plants
hindwings	the rear wings (closest to the tail)
host plant	the plant a butterfly or moth uses to lay its eggs on or caterpillars feed on

instar	a stage of development of a caterpillar or moth: most grow through several instars
iridescent	lustrous colors that can change depending on the light and angle of view
larva	the first stage of a butterfly's life after hatching from an egg: a caterpillar
lateral	sideways
light pollution	the amount of artificial light that invades an ecosystem at night
median line	the center
metamorphosis	the process of insects changing from larva to pupa to adult
migration	the seasonal movement of organisms, sometimes over several generations
mimicry	the technique of looking like a different organism
monoculture	when a single species (like turf grass) takes over and reduces biodiversity
morph	a different color or shaped version of the same species
nectar	the sugary fluid secreted by flowers to attract pollinators
neonicitinid	a targeted insecticide derived from nicotine
noxious	poisonous, toxic or unpleasant
osmetarium	an organ in the front of a caterpillar that can be everted when the caterpillar perceives a threat
pesticide	a poison meant to kill insect pests
photosynthesis	the ability of a plant to turn sunlight into energy
pollinator	an organism that collects or moves plant pollen from one flower to another, fertilizing the plant
proboscis	the tube-like tongue of a butterfly or moth
protuberance	a body part that bulges or sticks out
puddle	butterflies gather at shallow puddles, sometimes in groups, to ingest minerals from the soil
pupa, pupae	the stage of an insect between a larva and adult: a chrysalis or cocoon
roost	a place where organisms rest or sleep safely
salacylic	a compound derived from willows: can be toxic to other organisms
spinnaret	an organ that helps insects like spiders and caterpillars spin silk
thorax	the central part of an insect body which uses muscles to control wings
ubiquitous	found everywhere
unpalatable	bad-tasting, inedible

Index

Sources

Alabama Butterfly Atlas. https://alabama.butterflyatlas.usf.edu/ Species accounts.

Brock, J. P., & Kaufman, K. (2003). Butterflies of North America, Kaufman Focus Guides. Hillstar Editions Houghton Mifflin.

Burgess, L. (2018). Butterflies of South Carolina. Clemson University Home and Garden Information Center. https://hgic.clemson.edu/factsheet/butterflies-of-south-carolina/

Clemson University (n.d.). Butterfly Habitat and Butterfly Garden. Clemson University South Carolina Botanical Garden. https://www.clemson.edu/scbg/index.html

Dziedzic, B. (2023). Raising Butterflies and Moths in the Garden (2nd ed.). Firefly Books.

Enhancing Pollinator Habitat. South Carolina Wildlife Federation (2022, February 25).https://www.scwf.org/pollinator-habitat

Fairchild, Judy Drew and Sheridan-Wilson, Lori: (2012, February 1) Butterflies of Dewees Island Active Learning Press, ed 2

Sources

Griffin, Becky. (n.d.). About the Great Pollinator Census, Resources for Educators. University of Georgia Extension. https://gsepc.org/educators

Hall , D., & Butler, J. F. (n.d.). Featured Creatures Phaon Crescent. Entomology Department of the University of Florida.https://entnemdept.ufl.edu/creatures/bfly/phaon_crescent.htm

Hall, D. W., & Butler, J. F. (n.d.). Featured Creatures Eastern Pygmy Blue. Entomology Department of the University of Florida. https://entnemdept.ufl.edu/creatures/bfly/eastern_pigmy_blue.htm

Home Habitats: How native plants provide the landscapes that help birds and butterflies thrive. Audubon South Carolina. https://sc.audubon.org/news/home-habitats-how-native-plants-provide-landscapes-help-birds-and-butterflies-thrive

INaturalist. https://www.inaturalist.org/home

Know Your Native Pollinators: Phaon Crescent. Florida Wildflower Association. https://www.flawildflowers.org/know-your-native-pollinators-phaon-crescent/

Lindwall, C. (2022, February 25). Neonicotinoids 101: The Effects on Humans and Bees. NRDC Natural Resources Defense Council. https://www.nrdc.org/stories/neonicotinoids-101-effects-humans-and-bees

Macgregor, C. (n.d.). Fatal attraction: How street lights prevent moths from pollinating. The Conversation. https://theconversation.com/fatal-attraction-how-street-lights-prevent-moths-from-pollinating-60331

Marshall, S.. (2017). Insects Their Natural History and Diversity (2nd ed.). Firefly Books.

Melcher, K., & Griffin, B. (n.d.). Pollinator Garden Design Guide: Activities for Youth, Schools, and Beginning Designers. University of Georgia. https://extension.uga.edu/publications/detail.html?number=B1570

Native Plants of North America. Ladybird Johnson Wildflower Center. https://www.wildflower.org/plants-main

Plants for Pollinators South Carolina Wildlife Federation. https://www.scwf.org/plants-for-pollinators

Shepard, M., Ph.D, & Farnworth, E. G. (2014). Garden Insects of South Carolina. Lowcountry Biodiversity Foundation.

Shepherd, M., PhD, Farnworth, E. G., PhD, & McCullough, K. L., MS (2010). Common Insects and Spiders of the South Carolina Lowcountry. Lowcountry Biodiversity Foundation.

(n.d.). Sleepy Orange Abaeis nicippe. Butterflies and Moths of North America. https://www.butterfliesandmoths.org/species/Abaeis-nicippe

(n.d.). Spinx Moths Hawk Moths. Missouri Department of Conservation. https://mdc.mo.gov/discover-nature/field-guide/sphinx-moths-hawk-moths

Tallamy, D. (2021). The Nature of Oaks: The Rich Ecology of Our Most Essential Native Trees. Timber Press.

Taylor, M. (2020). How Insects Work: An Illustrated Guide to the Wonders of Form and Function. The Experiment.

Wagner, D. L. (2005). Caterpillars of Eastern North America, Princeton Field Guides. Princeton University Press.

Whalley, P. (1988). Butterfly & Moth. Doring Kindersley.

Wheeler, J. (n.d.). To Protect Moths—Turn Out The Lights! Xerces Society for Invertebrate Conservation. https://www.xerces.org/blog/to-protect-moths-turn-out-lights

Williams, W. (2020). The Language of Butterflies: How Thieves, Hoarders, Scientists, and Other Obsessives Unlocked the Secrets of the World's Favorite Insect. Simon and Schuster.

www.ingramcontent.com/pod-product-compliance
Lightning Source LLC
Chambersburg PA
CBHW041307020426

42333CB00001B/9